MODELS IN TECHNOLOGY AND SCIENCE

Shadow Play
Making Pictures with Light and Lenses

BERNIE ZUBROWSKI
Illustrated by Roy Doty

STUDENT'S GUIDE

Pitsco, Inc.
Pittsburg, Kansas

57844

To Olivia and Quatie

Library of Congress Cataloging-in-Publication Data
Zubrowski, Bernie
Mirrors: finding out about the properties of light / by Bernie Zubrowski; illustrated by Roy Doty.
p.   cm.
Summary: Suggested activities explore how mirrors work and how they demonstrate the properties of light.
ISBN 1-58651-900-X
1. Mirrors – Experiments – Juvenile literature. [1. Mirrors – Experiments. 2. Light – Experiments. 3. Experiments.] I. Doty, Roy, 1922- ill. II. Title.
QC385.5.Z83 1992
535'.078 – dc20   91-29142   CIP   AC

Acknowledgements

Thanks to Maurice Bazin, who checked the accuracy of the scientific content, and extra-special thanks to Marjorie Waters, who helped me put the final manuscript into clear and coherent form. Also to the students at the Hennigan School of Boston and the Russell Street School of Littleton, Massachusetts, who helped me try out the projects in this book.

Table of Contents

Introduction

People have been fascinated by shadows for a long time.

Some of the oldest pictures in the world are drawings in caves in France and Spain. Scientists estimate that they were done thousands of years ago. Most of these show individual figures rather than entire scenes. Since the drawings are in caves, it may be that these early artists might have started by making outlines of some of the shadows cast by torches and fires.

Making pictures has also been practiced by many artists throughout the world for a very long time. The content of the pictures has taken many different forms. Some are only outlines of objects. Some, such as those made by the aborigines of Australia, try to show all sides of a person or object at the same time. Others show scenes in which near and faraway objects appear close together, resulting in a picture very different from a photograph.

Starting around the thirteenth century, artists in Western Europe experimented with a number of primitive cameras that allowed them to produce pictures that were more realistic–sometimes looking like photographs. Not only did these explorations help them understand how to produce pictures that were miniature replicas of outdoor scenes, they also led to discoveries about some of the basic properties of light. In some ways these artists were the first scientists of light.

The next major development in picture making was the invention of the camera and of light-sensitive film. In order to produce pictures of good quality, photographers and scientists experimented with lenses and with different kinds of chemicals on film. These explorations furthered the understanding of the special properties of light.

These early cameras were very primitive. It took a long time for the chemicals on the film to react to the light source, so the person or object being photographed had to remain still for a while. Because the equipment was cumbersome and the film and chemicals for developing the film were expensive, only a few people were able to make photographs at first. Gradually, easy-to-use cameras and film were developed and everyone could take his or her own pictures.

This book shows you how to use everyday materials to carry out explorations like those done by artists and light scientists of the past. It is written in three parts. In the first part, you will experiment with making shadows in natural and artificial light. In the second part, you will make and use a shadow box to do more controlled experiments. And in the third part, you will make and use a box camera.

These explorations will help you understand some of the basic properties of light better – and will improve your drawing skills at the same time!

SAFETY NOTES:
You will be using a slide projector in some of the experiments in this book. Don't let the light shine directly in your eyes.

You will also be using a utility knife to cut out cardboard. Ask an adult to help you – and handle the knife carefully.

Casting Shadows

In this section, you will look closely at shadows, both outdoors in sunlight and indoors in artificial light. You will have to be a keen observer and pay special attention to the way light is blocked or bent to form a shadow. The experiments you do will help you observe how sunlight differs from artificial light. Some of the results may surprise you!

Shadows in Sunlight

When the sun shines, shadows are cast on the ground, creating an outline of any object that is upright. Shadows are so commonplace that you probably don't pay much attention to them. However, there is much to discover if you take the time to observe shadows and experiment with making your own. While it is easy to see shadows changing both their size and their shape, they have other characteristics that require a more careful examination. You will be examining some of these in this section.

To examine shadows in sunlight, you will need:

▼ Several pieces of chalk
▼ Broomstick or dowel, 3 feet long and at least 1/2 inch in diameter
▼ Cardboard tube, such as the kind that holds wrapping paper, or a tube made from
▼ Sheets of newspapers, 3 feet long and approximately 3 inches in diameter
▼ Triangular piece of cardboard, 10 inches on each side
▼ Square piece of cardboard, 10 inches on each side
▼ Round piece of cardboard, 10 inches in diameter
▼ Piece of Plexiglas or other transparent plastic, 12 inches square (This can be found in hardware and building-supply stores.)
▼ Piece of small-grid wire mesh, approximately 10 inches square (This is sold as window screening. It can be found in hardware stores.)
▼ Piece of large-grid wire mesh, approximately 10 inches square (This is sold as hardware cloth. It can be found in hardware stores.)
▼ 4 dowels, 10 inches long and 1/8 inch in diameter
▼ Rubber bands

Getting Started

Join the four 10-inch dowels together with the rubber bands. They should form a square frame.

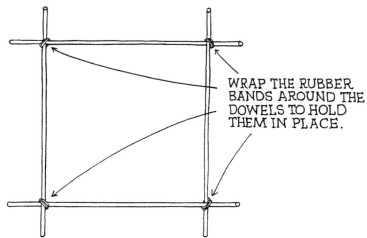

WRAP THE RUBBER BANDS AROUND THE DOWELS TO HOLD THEM IN PLACE.

Find a spot outside where the sun shines on a sidewalk or driveway. Carry all the objects on the list plus your dowel frame to the place you've chosen. The best time to explore shadows is midmorning or midafternoon, when the shadows are neither too long or too short. It is helpful to have a friend use a piece of chalk to draw the outline of the shadows. You should have a notebook handy to make drawings of some of the shadows, to describe how you made them, and to record any other interesting observations. When making drawings of your shadows, also include the time of day.

The overall purpose of these activities is to look closely at how shadows are formed in sunlight. Therefore, you should take note of how sizes and shapes change. Also pay attention to whether the edges of the shadows are sharp or fuzzy.

To begin, stand so that the sun is in back of you. Arrange your arms, legs, and head in different positions. Notice how the shape and size of the shadows change.

Next, hold up any object that is handy. Turn it in different positions. Notice the kinds of shadows that it makes as it rotates.

Can you make any general observations about all of these different shadows?

Experiments to Try

Now try the following experiments.

▼ What is the biggest shadow you can make with your own body? As you strike a pose or position your body, have your friend draw the outline of your shadow on the sidewalk. You and your friend can take turns trying different positions and outlining the shadows.

▼ What is the smallest shadow you can make with your body? How do you have to position yourself?

▼ What are the biggest and smallest shadows you can make with the broomstick or the 1/2-inch-diameter dowel? How do you have to position them?

▼ Can you line up the cardboard tube so that sunlight comes through the opening, forming a round patch of sunlight with a ring of shadow around it?

▼ How must you hold one of the cardboard shapes so that a shadow in the shape of a straight line appears on the sidewalk?

▼ What happens to the shadows of the triangular, square, and circular pieces of cardboard as they are rotated in different positions?

▼ Does the size of these shadows change as you move the shapes closer to or farther away from the ground?

▼ What kind of shadow do you get with the piece of Plexiglas? How does this shadow change as you rotate the piece of Plexiglas?

▼ What happens to the shadows of the window screening and the hardware cloth as you hold them close to the ground and them move them several feet up in the air?

▼ What happens to the shadow of the dowel frame as you move it close to and farther from the ground?

▼ Make other cardboard shapes and other kinds of frames. Try them in different positions and at different distances from the ground. See what different kinds of shadows you can make.

Remember to record your observations and drawings in your journal.

What's Happening?

If you were paying close attention to how you lined up your body and the objects, you should have noticed that the size of the shadow depends on whether you stood up straight or leaned toward or away from the sun.

Shadows cast by the broomstick and the cardboard tube are the most revealing. You can make a small dot on the ground with the broomstick and the cardboard tube by pointing one of the ends right at the sun.

You can make your body's shadow as small as possible by pointing yourself at the sun – you may have to lean on a friend. The same thing happens when you line up the smallest side of the cardboard shapes or any other object with the direction of the incoming sunlight; you will make the smallest possible shadow. On the other hand, to make the biggest shadow, you have to turn the object full face to the sunlight.

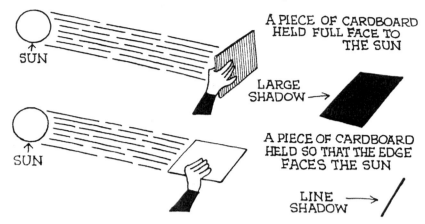

One way of thinking about sunlight is to compare it to rain falling from clouds. If the wind is not blowing, rain falls straight down. In this situation you hold your umbrella straight over your head to keep dry. During other kinds of rainstorms the wind may be blowing hard, causing the drops to fall at a slant. Then you have to hold your umbrella at a slant so you don't get wet.

Now, think of light coming from the sky as very tiny particles streaming to Earth. Wind is not affecting these particles, because they are like packets of energy. Depending on the time of day, they are either streaming at an angle to Earth or coming from almost straight overhead. At noontime, the sun is almost directly overhead. Therefore, the light particles are streaming to the earth straight down. (The sun is only *directly* overhead near midday in the tropical regions of Earth.) If you place your body, a stick, or a thin object in a vertical perpendicular position, you will make the smallest shadow you can make, because you are exposing the smallest surface to the stream of light particles. In the morning or afternoon, the light particles are streaming to Earth at an angle. If your body or an object is standing straight up, it will be blocking the stream of light. Therefore, the shadow will be bigger. The lower the sun is in the sky, the more your body or an object will be blocking the light stream and the bigger the shadow will be. But if you lean your body or point a pole directly toward the sun, you decrease the surface that is exposed to the stream of particles, and therefore make a smaller shadow.

Thinking of light as being composed of very tiny particles helps to explain some of the results you have gotten. But that is not enough to explain the results of other explorations. For instance, light acts in peculiar ways when it goes around the edges of objects, when it moves past thin materials, or when it travels through small holes.

The shadows from thin objects such as window screening or a thin dowel need to be considered very carefully. When these things are placed close to the ground, you do get thin dark lines, but these start to fade as you move the object farther from the earth. The shadows become fuzzy gray lines. Eventually, there seem to be no shadows at all.

These results would seem to be contrary to the idea that sunlight reaching the Earth is a stream of parallel rays. If they were perfectly parallel, then there should be a shadow of the thin wire even when it is several feet from the ground.

To better understand what is going on, it is helpful to think about what happens when you place your hand near a wall in a room where there are two glowing lamps. When your hand is a certain distance from the wall, you get two kinds of shadows. The middle area is dark, while the sides are gray. Scientists call the dark area the *umbra* of the shadow, while the gray area is called the *penumbra*.

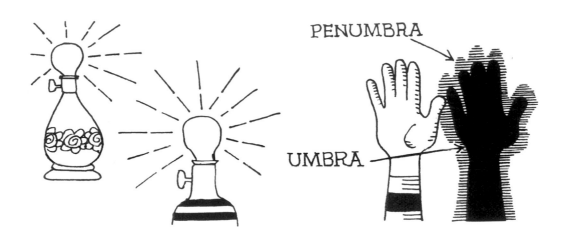

You can see how this occurs by tracing light rays from the two light sources. The light from one light source is coming at a different angle to the object that the light rays from the other light source. The rays from one source go underneath the object on one side, while the rays from the other source go underneath the object on the other side. The dark area is where no rays of light reach.

Now, consider light coming from the sun. You can think of light coming from the sun in a similar way. Some light shines a little underneath one side of the object, and at the same time some light shines underneath the same object on the other side. You then have a shadow with an umbra and a penumbra. The closer you move an object to the ground, the less light shines underneath the object, making the umbra larger. You can reach a point very close to the surface of the ground where it appears that there is no penumbra at all. If you move the object farther and farther away from the ground, the umbra of the shadow disappears and the penumbra becomes larger.

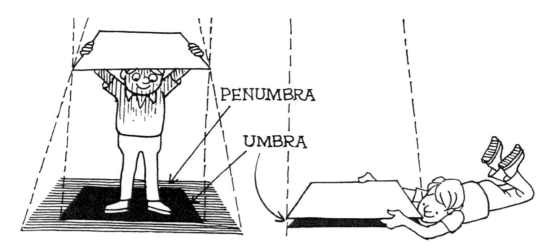

From your experiments, you should have seen that the Plexiglas itself doesn't make a shadow but that the edges and the scratches on the surface do. They show up on the ground as curved or straight lines. Later explorations will look at transparent objects again to see what happens when they are curved or of varying thicknesses. The results are very special and will help you understand what happens with light and lenses.

You have seen that light is blocked whenever it hits an opaque object. (*Opaque* means that the object is neither reflecting light nor letting it pass through.) What is not readily apparent is why shadows have the shape and size that they do. In sunlight, if you are very careful in positioning the object, the shadow is very close to the size of the object. Also, if you move it away from the ground while keeping it at the same angle to the ground, the size of the shadow will hardly change.

You should also have seen that in many positions, the shadow doesn't look like the object at all. The next set of activities will help you see how this happens and give you a better sense of how these different shapes occur.

~~~~~~~~~~~~~~~~~~~~~~

Shadows in Artificial Light

For a long time the only sources of light were the sun or burning wood or oil. Candles are a recent invention, and the incandescent lightbulb has only been around for a hundred years. Some of these artificial light sources give off dim light, while others are very bright. An important question is whether the light coming from these sources will make shadows in the same way as the light coming from the sun.

Is artificial light like sunlight? Does it make shadows in the same way? To begin to answer these questions, you will first repeat the experiments from the last section, using artificial light instead of sunlight. Remember to make drawings and take notes of what you observe.

You will need:
▼ Materials from page 9
▼ Tracing paper
▼ Masking tape
▼ Light source, preferably a slide projector (or you can use a table lamp with a bare incandescent lightbulb that is clear – not frosted – and cylindrical)
▼ Projection screen, large white bed sheet, or large piece of white paper (optional)

Getting Started

You will need a big room that can be darkened. Find the longest and widest space so that the projector can be set up as far from the wall as possible. A plain wall without decorations is needed in order to see the shadows clearly. You can hang a large white bed sheet or a large piece of white paper on the wall, or put up a projection screen.

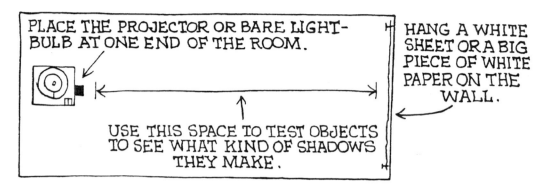

PLACE THE PROJECTOR OR BARE LIGHT-BULB AT ONE END OF THE ROOM.

USE THIS SPACE TO TEST OBJECTS TO SEE WHAT KIND OF SHADOWS THEY MAKE.

HANG A WHITE SHEET OR A BIG PIECE OF WHITE PAPER ON THE WALL.

Place all kinds of different objects in the beam of light. Try to use objects that you think will make interesting shadows. Make drawings of their shadows, and think about the ways your results are the same or different from the results you obtained in sunlight.

Next, tape tracing paper to the wall. Position your head in such a way that the shadow of your profile falls on the paper. Have a friend trace the outline, then do the same for your friend. These outlines are called *silhouettes*. If you have large sheets of paper, you can also make silhouettes of your whole body.

Experiments to Try

Repeat the experiments from pages 10–11.

▼ What happens to the shadows as you move the objects closer to or farther away from the light source?

▼ What happens to the edges of the shadows? Are they sharp or fuzzy?

▼ Make a pencil hole in a cardboard shape. Hold the cardboard in front of the projector so that a very small beam of light is projected onto the wall or screen.

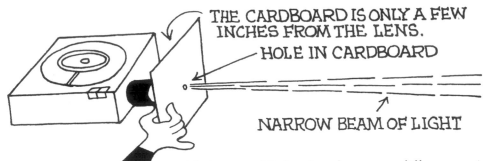

THE CARDBOARD IS ONLY A FEW INCHES FROM THE LENS.

HOLE IN CARDBOARD

NARROW BEAM OF LIGHT

Try all of the objects again in this beam of light. Are there any differences in what you observe about the size or sharpness of the shadows?

▼ Stand in the middle of the light beam, about six feet from the wall. Try to make the smallest possible shadow with the dowel or broomstick. Then move toward the right edge of the light beam, keeping the same distance from the wall. Try to make the smallest possible shadow with the dowel. Move to several other positions, keeping the dowel six feet from the wall. Each time, try to make the smallest possible shadow.

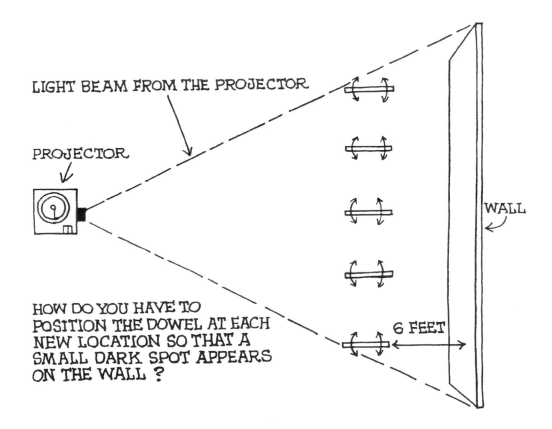

LIGHT BEAM FROM THE PROJECTOR

PROJECTOR

WALL

HOW DO YOU HAVE TO POSITION THE DOWEL AT EACH NEW LOCATION SO THAT A SMALL DARK SPOT APPEARS ON THE WALL ?

6 FEET

▼ Do the same thing using the hollow cardboard tube. Do you have to change the angle of the tube as you move it farther from the light?

What's Happening?

Most of the results that you obtain are similar to those you saw in sunlight. The closer you are to the wall, the sharper the shadow. However, there is one major difference. In sunlight the size of a shadow changes very little as you move an object closer to or farther from the ground. In artificial light, the closer the object is to the wall, the smaller its shadow.

The experiments with the dowel and the cardboard tube show another major difference between natural and artificial light. In order to make a small shadow, you have to point the object at the projector as you move to different positions along the wall.

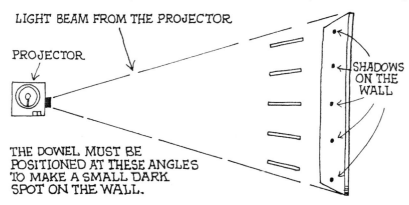

There is a reason for these differences. Light from an artificial source spreads out, or *diverges*, as it travels away from the source, even when you put a piece of cardboard with a pencil hole right in front of the light to make the beam smaller. In contrast, because the sun is so far away, sunlight reaches the Earth in straight parallel lines called *rays*. You can't actually see the rays, but scientists use lines to show the direction in which the light is traveling. The lines show how most of the light moves from a source and what happens when it lands on an object.

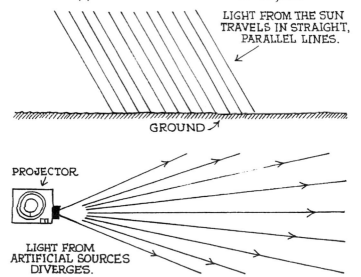

Even though sunlight and artificial light behave differently, the light itself is the same no matter what the source. Light is a form of energy that at times can be thought of as tiny particles shooting out of a source.

There is another important difference between artificial light and sunlight. Shadows from bare incandescent bulbs are usually fuzzy unless the objects are very close to the wall. When a lens is used, or when the light travels through a pencil hole in a piece of cardboard, the shadows are sharper and objects can be farther away from the wall before they get fuzzy. (You will be doing more experiments with lenses and holes in later activities to help you understand how this happens.)

The shadows of the screens and the dowel frame change in the same way in artificial light as they did in sunlight. The farther these objects are from the wall, the fuzzier the images. Here, too, the shadows of the individual dowels almost disappear as you move far away from the wall.

The shadow of the piece of Plexiglas is a little different in artificial light from its shadow in sunlight. The scratches will be more apparent on the wall when a projector light is used. (More experiments with transparent objects and shadows are coming up in the next activities.)

The major observation you should make from the explorations in this section is that light diverges from the filament of a lightbulb. (A *filament* is the wire in the lightbulb.) If the filament is small, it can be considered a point source of light. This kind of source makes the best shadows. In this situation the light comes off the filament of the lightbulb in every direction in straight lines. A shadow results when some of the rays are blocked by the object while some continue toward the screen.

Further Explorations

You can make the light from an artificial source travel in almost the same way as sunlight. Using an arrangement of mirrors, you can create a situation where the size of the shadow doesn't change very much as you move an object close to or farther away from a wall.

You will need:

▼ Slide projector
▼ 3 identical mirrors, each at least 4 inches wide (Large-size mirrors can be found in the cosmetic sections of large department stores.)
▼ 4 straight-backed chairs of equal height
▼ Masking tape, 1-inch wide
▼ Large piece of paper

You will need a large, very dark room. Set up the projector on a chair in one corner so that the light shines on the opposite wall. Position Mirror #1 on another chair about six feet from the projector so that the light beam hits Mirror #1 and bounces off in a different direction. About six feet from Mirror #1, position Mirror #2 so that it reflects light (or throws it off at an angle) from Mirror #1. Then position Mirror #3 about six feet away from Mirror #2 so that it *reflects* light from Mirror #2 onto a plain white surface, such as a wall or a large piece of paper. All the mirrors must stand up straight, perpendicular to the floor.

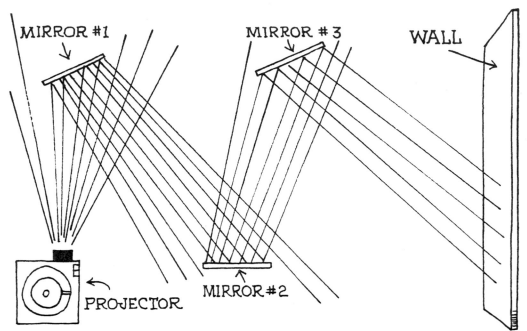

The light from Mirror #3 will be faint, but it should still be able to cast a light on the wall. Put an object such as a small square piece of cardboard in the light beam so that it is parallel to Mirror #3. Slowly move the object closer to the wall, then closer to the mirror. Watch carefully to see if there is a noticeable change in the size of the shadow. Also, look carefully at the edge of the shadow.

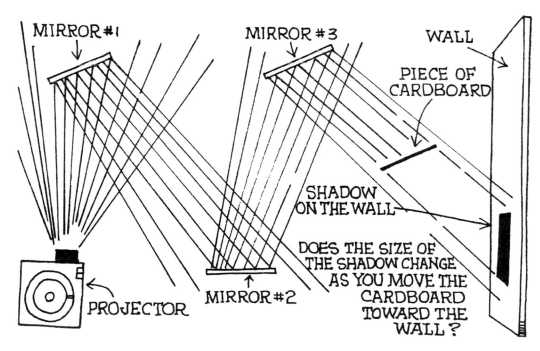

As you move the piece of cardboard away from the wall toward Mirror #3, the size of the shadow will not change very much, especially when compared to the size changes in the previous activity.

You can picture in your mind what happens to the light beam by studying the drawing above. As you have already discovered, the light spreads out when it leaves the lens of the projector. Mirror #1 reflects only part of this light beam. The broadest rays move on past the mirror, so the reflected light is not as spread out as the original beam coming from the lens. Mirrors #2 and #3 cut down the spreading even more. By the time the light is reflected off Mirror #3, the rays are almost parallel to one another, like rays of sunlight.

Shadow Box

It is not always convenient to experiment with shadows outdoors because the sun doesn't always shine. And in a room indoors, furniture may get in the way of the light beam. One solution to these problems is to make a small laboratory, called a *shadow box*, where you can experiment with shadows in different ways without any outside interference.

Making and Using a Shadow Box

This activity will show you how to convert a cardboard box into a shadow box and test different light sources.

Making a Shadow Box

You will need:

▼ Cardboard box, approximately 12 inches wide, 12 inches high, and 24 inches long, or two smaller boxes that are the same size (the exact dimensions are not critical.)
▼ Piece of Plexiglas or other transparent plastic, approximately 1/8 inch thick, 8 inches wide, and 11 inches long
▼ Scissors or utility knife
▼ Pencil
▼ Duct tape, at least 2 inches wide
▼ 2 large paper clips
▼ Pad of tracing paper, 9 inches wide and 12 inches long
▼ Table lamp with bare incandescent lightbulb that is clear – not frosted – and cylindrical
▼ Slide projector

Step 1. If you are using two smaller boxes, place them side by side to make the longest box possible. Cut off the sides where the boxes meet. Tape the two small boxes together so that you have a long box.

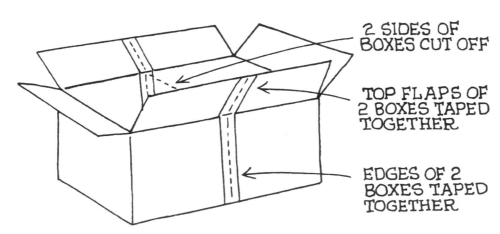

2 SIDES OF BOXES CUT OFF

TOP FLAPS OF 2 BOXES TAPED TOGETHER

EDGES OF 2 BOXES TAPED TOGETHER

Step 2. Place the piece of Plexiglas on one of the narrow sides of the box. Line up the Plexiglas so that all its sides are the same distance from the sides of the box.

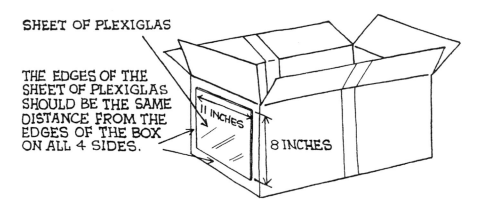

SHEET OF PLEXIGLAS

THE EDGES OF THE SHEET OF PLEXIGLAS SHOULD BE THE SAME DISTANCE FROM THE EDGES OF THE BOX ON ALL 4 SIDES.

11 INCHES

8 INCHES

With a pencil, trace the outline of the Plexiglas. Carefully cut out this section to form a window. Do not cut off the top flaps of the box.

Step 3. Place the piece of Plexiglas in the window that you have cut out. Tape the two sides and the bottom of the piece of Plexiglas to the box with duct tape. (Do not tape the top side.) Tape just the edges of the Plexiglas so that most of the piece is clear.

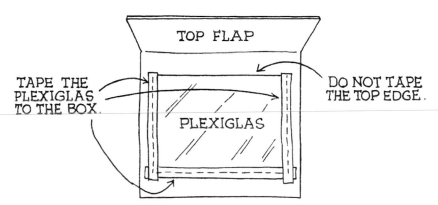

TOP FLAP

TAPE THE PLEXIGLAS TO THE BOX.

DO NOT TAPE THE TOP EDGE.

PLEXIGLAS

Step 4. Slide two large paper clips onto the top edge of the Plexiglas. Then slide a piece of tracing paper beneath the paper clips so that the paper covers the piece of Plexiglas. This will be your screen.

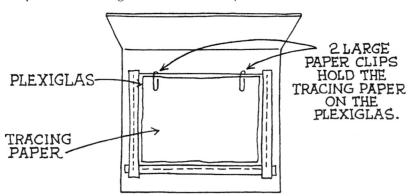

PLEXIGLAS

TRACING PAPER

2 LARGE PAPER CLIPS HOLD THE TRACING PAPER ON THE PLEXIGLAS.

Making a Bulb Holder

You will need:

▼ 7.2-volt flashlight bulb
▼ Thin-wall plastic tubing, 1-inch long by 1/2-inch in diameter (This is usually sold by the foot at most hardware stores.)
▼ Nail, two inches long
▼ 2 pieces of electrical wire, each about 12 inches long (This is sold by the roll at most hardware stores or at Radio Shack.)
▼ D-size battery
▼ Scissors or knife

Step 1. Using a knife or scissors, cut a one-inch piece of plastic tubing. Force the nail through the middle of this piece.

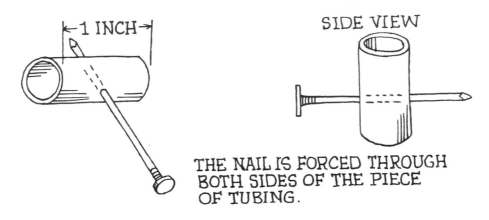

THE NAIL IS FORCED THROUGH BOTH SIDES OF THE PIECE OF TUBING.

Step 2. Wrap one end of a bare electrical wire around the head of the nail.

Step 3. Place a bare end of the other electrical wire inside the tubing so that the wire is flush against the plastic but not touching the nail. Slide the metal base of the flashlight bulb into the tubing until it touches the nail. (The base of the bulb should hold the second wire in place against the tubing.)

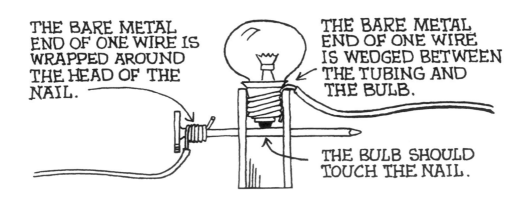

THE BARE METAL END OF ONE WIRE IS WRAPPED AROUND THE HEAD OF THE NAIL.

THE BARE METAL END OF ONE WIRE IS WEDGED BETWEEN THE TUBING AND THE BULB.

THE BULB SHOULD TOUCH THE NAIL.

To test your completed bulb holder, touch the free ends of the two electrical wires to the two ends of a D battery. The bulb should light. If the bulb does not light, check all the connections to make sure the metal parts are touching each other.

Getting Started

You can use this shadow box to experiment with different light sources to see if they create images of the same size and shape.

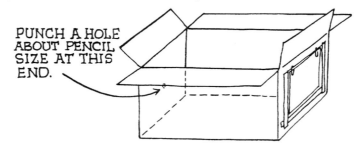

PUNCH A HOLE ABOUT PENCIL SIZE AT THIS END.

Before you begin experimenting, punch a hole with the pencil in the middle of the side of the box opposite from the piece of Plexiglas.
[graphic]

Try to predict what image you will see on the Plexiglas screen if you shine a light through the hole.

Then take the shadow box outdoors and hold it so that the hole in the box faces the sun directly, as shown in the drawing. What will be the size and shape of the image the sunlight casts on the Plexiglas screen? Is the image bigger, smaller, or the same size as the hole in the box?

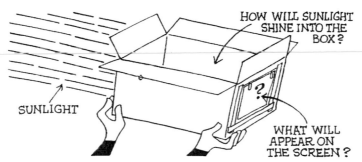

HOW WILL SUNLIGHT SHINE INTO THE BOX?

SUNLIGHT

WHAT WILL APPEAR ON THE SCREEN?

Experiments to Try

Carry out the same investigation indoors with the table lamp and the bare incandescent light. Place the lightbulb next to the hole, outside the shadow box, and observe how light shines on the Plexiglas screen.

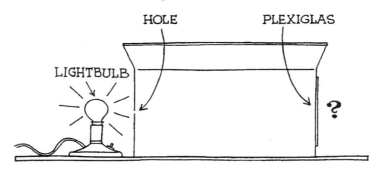

HOLE   PLEXIGLAS

LIGHTBULB

?

▼ Move the lightbulb three feet from the hole. Now what happens to the light as it shines through the hole in the shadow box?

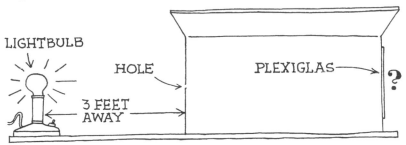

▼ Try to predict what will happen if you use the slide projector and flashlight bulb. Then try each of these light sources, first close to the box, then three feet away. What images appear on the screen?

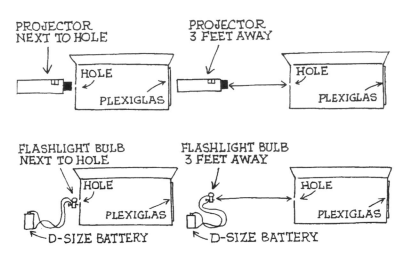

What's Happening?

In the sunlight test, only a small spot of light shows up on the piece of Plexiglas. This is because the light rays from the sun travel in parallel lines. Since they don't spread out, they cast a narrow beam of light.

You get similar images from these three artificial light sources when they are placed three feet from the shadow box because the hole allows only some of the light to reach the Plexiglas screen. The hole also narrows the beam of light, so the image on the Plexiglas screen is a small circle with a clear, focused edge.

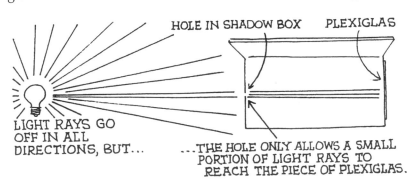

But the three light sources do not shed light in the same way, as you can see when you put them right next to the hole. The incandescent lightbulb and flashlight bulb light up the entire Plexiglas screen. The slide projector produces a circle of light that does not quite fill the whole Plexiglas screen.

All three light sources give sharp-edged shadows, but those from the incandescent lightbulb are a little fuzzier than those from the other two sources. Since the light from the small flashlight bulb gives the best shadows for your experiments, it will be used for the activities that follow.

Shadows of Shapes

The shadow box can be used to test objects of different shapes and sizes. Because it has tracing paper at one end, you can see the shadow outside the box. This arrangement will allow you to trace outlines of the shadows to make drawings of the objects. You can use these drawings to play guessing games with your friends.

To play the shadow game, one of you holds an object in the shadow box. The others try to guess from the shadow alone what the object is.

Shadow Games: Three-Dimensional Shapes

In the first shadow game, the objects are three-dimensional.

You will need:
▼ Shadow box from pages 23–24
▼ Bulb holder from page 25
▼ Battery holder for 4 D-size batteries (This is available at Radio Shack or a hobby shop.)

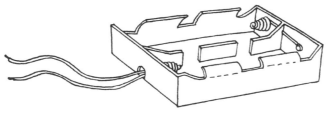

▼ 4 D-size batteries
▼ Package of oil-based clay, such as Plasticine
▼ Several pieces of thin, stiff, straight wire, such as florist wire, each approximately 5 inches long

Step 1. Set up the shadow box in a dark room. Put four D-size batteries in the battery holder. Connect the free ends of the two pieces of electrical wire on the bulb holder to the battery holder, as shown. The bulb should light.

2 WIRES FROM THE BULB HOLDER CONNECTED TO 2 WIRES FROM THE BATTERY HOLDER.

Step 2. Insert the flashlight bulb into the small hole in the shadow box.

Getting Started

To get a sense of how three-dimensional objects make shadows, spend some time playing with familiar small objects. For example, hold a toy in the shadow box so that a shadow appears on the Plexiglas screen. Can your friends guess what the toy is by the size and shape of the shadow? Does rotating the object help them guess correctly?

Next, try making different shapes with the Plasticine. Form as many interesting ones as you can think of. As you do this, make drawings of the more interesting shadows and the shapes that created them. What properties to the different shadows seem to have in common?

Experiments to Try

Mold the Plasticine into four shapes: a cube, a tetrahedron, a sphere, and a cylinder. Then insert a piece of wire into each piece of Plasticine so that you can hold the wire and rotate the object.

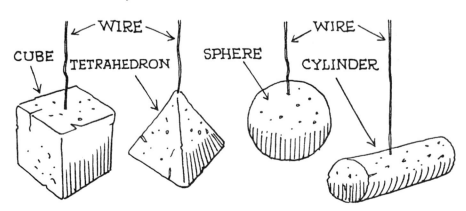

CUBE   WIRE   TETRAHEDRON   SPHERE   WIRE   CYLINDER

▼ Can you make the shadow of the tetrahedron look triangular?

▼ Can you make the shadow of the cube look square?

▼ Can you make the shadow of the sphere look circular?

▼ Can you make the shadow of the cylinder look rectangular? Circular?

▼ How does the shadow change as you move each of these shapes to different parts of the box?

▼ How does the shadow change as you rotate each piece?

▼ Can you guess how large an object is by the fuzziness of the shadow?

What's Happening

Regardless of what object you use, the shadows are similar to those you saw when you set up the slide projector in a dark room for the experiments on pages 16–17. Moving the object closer to the light makes the shadow bigger and fuzzier. Moving it closer to the Plexiglas screen makes the shadow smaller and sharper. If you hold the object near the side of the box, the shadow may appear on the side of the box instead of the Plexiglas screen. This is because of the divergence of the light beams.

Rotating the shapes to a position where their faces are parallel with the Plexiglas screen will produce regular geometric shapes. It doesn't matter how you hold the sphere; you will always get a circular shadow shape. You can rotate the cube six different ways to get squares, and the tetrahedron can be rotated three ways to get triangles. No matter how you move any of these shapes you never get thin lines; you get dark, full shadows. (You should keep these observations in mind when you try out the next activities. The same shapes will be used, but they will be of different dimensions.)

As you observed with sunlight and the projector light, the edges of an object's shadow become fuzzier the farther away the object is from a wall or screen. In the last experiment, the distance between the light source and the surface on which the shadow is falling is short compared to the two previous experiments. Shadows on the screen of the box are the sharpest when the object is right next to or only a few inches away from the screen. When an object is near the lightbulb, the edges of its shadow are fuzzy. This information is useful if you have no idea of the size of the object inside the box. A cube four inches on a side might have the same size shadow as a cube one inch on each side – if the smaller cube is held farther away from the screen. By examining the edge of the shadow, you can guess whether the object is a four-inch or a one-inch cube.

At this point you should think about the difference between a real object and its shadow. The shadow shows the outline of the object, but only from one side. If the object is symmetrical, like the cube or the sphere, it doesn't make any difference which side produces the shadow. However, most objects need to be rotated before you can know from the shadow alone what the object is. This means that you need to present several different sides of an object to know it fully. For this reason, engineering drawings of a machine or tool usually show top, bottom, and side views so that someone can look at the drawing and build the object correctly.

Further Explorations

You can do an interesting art project by adding more holes to the end of the shadow box and putting a light source in each one. You will then get multiple shadows of the same object.

Follow the directions for making bulb holders from page 25. Punch four or six holes in one end of the shadow box, equidistant from one another. Place a flashlight bulb in each hole. Connect the wires from the bulb holders to the battery holder. (See page 30 for assembly directions.)

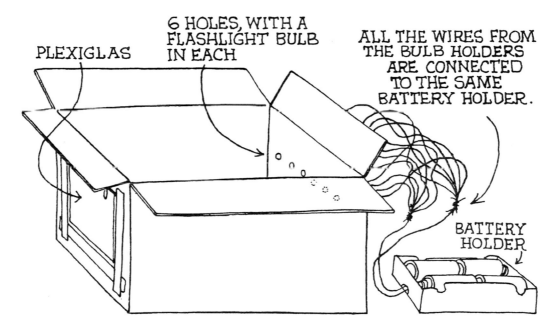

One at a time, put different objects inside the shadow box. Rotate each one and watch the multiple shadows on the screen.

When you're finished, remove the extra flashlight bulbs and cover the holes with duct tape so that your shadow box is back to its original form, with one hole only. Then you can use the same shadow box for the next activities.

Shadow Games: Two-Dimensional Shapes

The objects you used in the first guessing game were three-dimensional. In this shadow game, you'll experiment with two-dimensional versions of these shapes to see what kinds of shadows result. Some shapes make the same kinds of shadows as their three-dimensional relatives. Others, however, cast different-shaped shadows depending on how they are rotated. For instance, you found that the sphere always makes a circular shadow no matter how you hold it. A flat circle, however, can make other shadow shapes.

Shadows are useful because their behavior is predictable. Artists use shadows to show realistically how objects look at odd angles. Scientists use them to determine the direction and kind of light rays that are coming from a light source. And mathematicians can write precise formulas based on how shadows behave because of their predictable properties.

You will need:

▼ Shadow box and flashlight bulb arrangement from pages 29–30
▼ Package of 3" x 5" index cards
▼ Scissors
▼ Pencil
▼ Several pieces of thin, stiff, straight wire, such as florist wire, each approximately 5 inches long
▼ Masking tape
▼ Hair comb
▼ Piece of small-grid wire mesh, approximately 6 inches square (See page 9 for suggestions on where to purchase small-grid wire mesh.)

Getting Started

In this guessing game, you and your friends draw flat mystery shapes on index cards, cut them out, and tape a piece of wire to each shape. Then as you rotate each shape in the shadow box, see if your friends can identify the shape from its shadow alone. The guessers can ask you to rotate the object clockwise, to move it toward the light, or to move it in any other way they think might help.

Before you do the experiments suggested here, first create lots of different shapes of your own. Do they all make shadows in the same way? Don't forget to record your observations.

Experiments to Try

Cut out geometric shapes of a circle, triangle, and square. They should be about the same size as the sphere, tetrahedron, and cube you made on page 30.

▼ Can you move each of these shapes in the shadow box so that their shadows are only a line?
▼ Can a round shape cast an oval shadow?
▼ Can a square shape cast a rectangular shadow?
▼ What happens to these shapes as you move them to different locations in the shadow box?
▼ What kind of shadow do you get when you hold a comb in the shadow box?
▼ Hold the comb so that the tips of the teeth touch the bottom of the shadow box. What happens when you move it close to or farther away from the light? What happens when you keep it in one place and rotate it?
▼ What kind of shadow do you get with the window screening?
▼ Try other flat objects around your house or school. Before you put them in the shadow box, try to predict what kind of shadow they will make.
▼ Place one of the shapes a few inches from the Plexiglas screen and rotate it until a line shadow is formed. What happens to the shadow if you hold the shape the same distance from the screen but move it from one side of the box to the other?

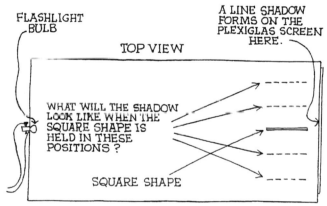

What's Happening?

The shadow will match the shape if you hold the shape parallel to the screen. Moving the shape closer to the screen will make the shadow smaller. Moving it closer to the light will make the shadow bigger.

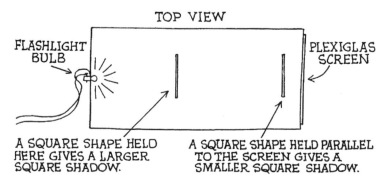

As soon as you start rotating the shape, the shadow will change. Rotating a flat circle will produce an oval shadow. Rotating a square will make a rectangular shadow. Rotating a triangle will make a skinnier triangular shadow.

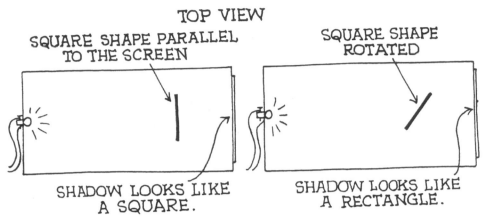

TOP VIEW

SQUARE SHAPE PARALLEL TO THE SCREEN

SQUARE SHAPE ROTATED

SHADOW LOOKS LIKE A SQUARE.

SHADOW LOOKS LIKE A RECTANGLE.

If you keep rotating a shape, you reach a point where there is only a line shadow on the Plexiglas screen. If you hold the shape parallel to the bottom of the shadow box, the line will be horizontal. If you hold it upright, the shadow will be vertical.

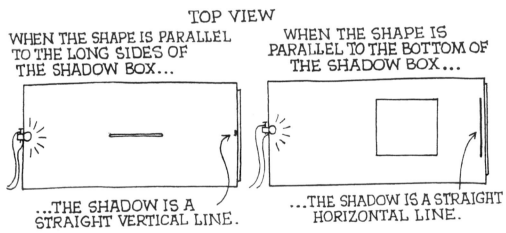

TOP VIEW

WHEN THE SHAPE IS PARALLEL TO THE LONG SIDES OF THE SHADOW BOX...

WHEN THE SHAPE IS PARALLEL TO THE BOTTOM OF THE SHADOW BOX...

...THE SHADOW IS A STRAIGHT VERTICAL LINE.

...THE SHADOW IS A STRAIGHT HORIZONTAL LINE.

These results are similar to those you observed in the experiments on pages 10–11 and 17, where you held a dowel or cardboard tube in sunlight or in the projector light. Those objects could be positioned so that they made the narrowest shadows.

When you experimented with the cardboard shapes in the projector light on page 16, you found that they had to be tilted slightly as they were moved from left to right to maintain the narrowest shadow. The same adjustment has to be made in this activity. To maintain the line shadow, you have to change the angle at which you hold the shape as you move it from left to right. This observation again shows that the light coming from the flashlight bulb is spreading out from the filament, and the object must be parallel to the direction of the light rays in order to cast the narrowest shadow.

Shadows produced by combs are very interesting and informative. Holding the comb parallel and close to the flashlight bulb will make divergent line shadows, as shown below. Moving the comb closer to the Plexiglas screen will make the shadows of its teeth almost parallel.

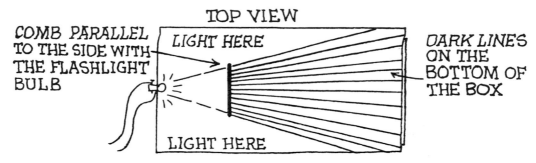

As you continue to rotate the comb the light beam has a narrower and narrower slit to go through. Therefore, the shadows of the teeth get thicker and thicker until no light appears between them, and you can't tell that the object is a comb.

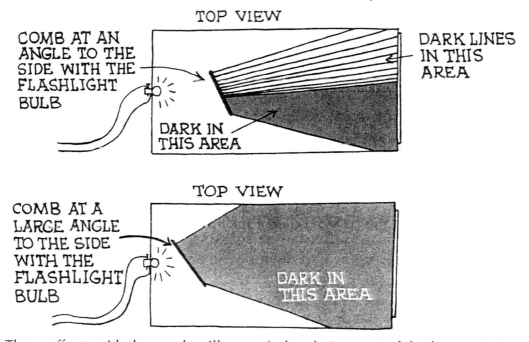

These effects with the comb will come in handy in some of the later activities. The shadow of the window screening that is made by the lightbulb in the shadow box is different from the shadows made by sunlight and the slide projector. As in previous situations, there is a good sharp shadow when the screening is close to the Plexiglas screen. Moving it farther and farther away from the screen does change the sharpness of the line, but not in the way it does with the other light sources. There are still dark lines instead of gray fuzzy ones.

Notice the shadow that the screening makes on the bottom of the shadow box when you hold the screening near the lightbulb. The shadow lines diverge from the surface of the window screening. This shows how the light itself is spreading out from the light bulb. There will also be lines on the Plexiglas screen as you move the window screening back and forth. The distance between the lines becomes farther apart as you move the window screening closer to the light bulb.

TOP VIEW

WINDOW SCREENING

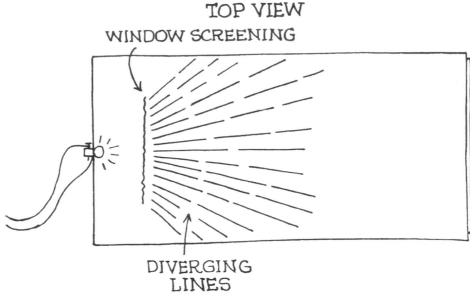

DIVERGING
LINES

When the window screening is near the Plexiglas screen, the shadow lines on the bottom of the box are almost parallel to one another, and the shadow on the Plexiglas screen looks like the window screening. Therefore, these results are similar to those obtained with the comb.

TOP VIEW

WINDOW SCREENING

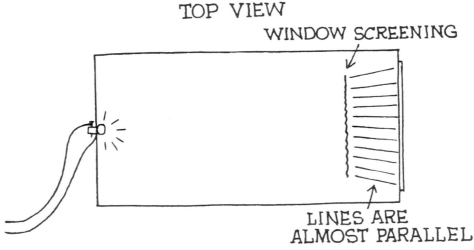

LINES ARE
ALMOST PARALLEL

The window screening and the comb results show directly how the rays of light are traveling from the filament of the flashlight bulb. As the rays go farther and farther from the light, they become more and more parallel. Remember the sunlight experiments in the first activity? The size of the shadow didn't change as an object was moved away from the ground because the sun is millions of miles away from Earth. By the time sunlight reaches the earth, its rays are traveling in parallel lines.

Shadow Games: Wire Shapes

You can go one step further in your experiments in reducing the number of dimensions of the objects that you are using to cast shadows. The results of experiments with wire shapes are quite similar to those in earlier activities, but you see outlines of objects instead of dark shadows. The shadow lines help you see the pathway of the light rays. Playing around with various wire shapes can also help you learn to draw outlines of objects.

You will need:
▼ Shadow box and flashlight bulb arrangement from pages 29–30
▼ 4 or 5 pieces of thin, stiff, straight wire, such as florist wire, each approximately 10 inches long
▼ Very thin string or sewing thread

Getting Started
As you did in previous activities, you should try making up your own shapes and experiments first. Keep notes of the interesting shapes and unexpected results you obtain.

Experiments to Try
▼ Bend one piece of wire to form a circle. Shape two other pieces of wire into a triangle and a square. What happens to the shadows of these shapes as you move them close to and far away from the light?

EACH SHAPE IS FLAT.

▼ What happens to the shadows of these shapes as you hold them in one place and rotate them?
▼ Make three-dimensional wire shapes of a cube, a tetrahedron, and a cylinder.

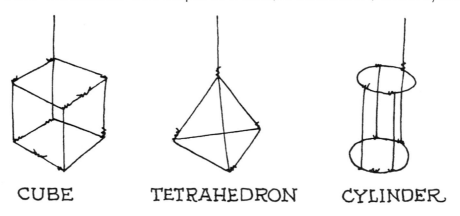

CUBE TETRAHEDRON CYLINDER

Make shadows with these shapes. At what point do they look like two-dimensional shapes? At what point do they look like three-dimensional shapes?

▼ Unwind one corner of a shape so that a leg hangs sideways. Can you still make the shadow look like the original shape?

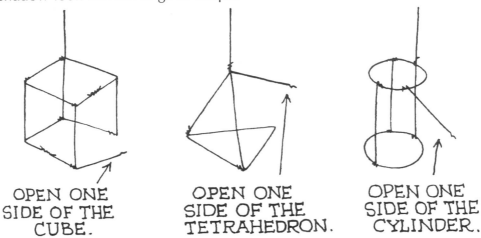

OPEN ONE SIDE OF THE CUBE.

OPEN ONE SIDE OF THE TETRAHEDRON.

OPEN ONE SIDE OF THE CYLINDER.

▼ Tie a piece of very thin string or sewing thread to the middle of a four-inch length of straight wire. Suspend the wire by the string in the box. What is the smallest shadow you can make with the wire? What is its shape?

▼ As you move the suspended wire from side to side, how do you have to adjust its position to maintain the smallest shadow?

What's Happening?

The results of your experiments with wire shapes are similar to the results of the two previous sets of experiments: In all three sets of shadows you can see the edges of the objects. The major difference is that with wire shapes you see an outline instead of a solid shape. This type of shadow gives you a little more information about what the object is really like. For example, compare the shadow of the flat, square, index-card shape from pages 33–34 to that of the wire cube. The shadow of the wire cube shows more lines, indicating that the object is three-dimensional. The same holds true for the tetrahedron and the cylinder.

Now compare the wire-shape shadows to the shadows made by the Plasticine shapes, on pages 30–31. In some positions you can't tell what the Plasticine shape is. The wire shape gives you more information.

Three-dimensional wire shapes give more complex shadows than ones made with flat shapes. However, the extra lines can help you better decide how the object is oriented. No matter how you rotate and twist the three-dimensional wire shapes, the shadows will always be of all the sections of the shape. This is because of the diverging beam of light.

You can make a very small spot on the Plexiglas screen with the straight piece of wire. (You have to ignore the shadow of the string the wire is suspended from.) As you move this wire to different locations, its position relative to the light changes.

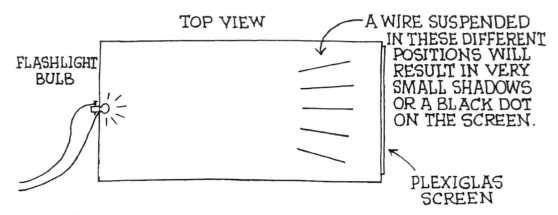

TOP VIEW

FLASHLIGHT BULB

A WIRE SUSPENDED IN THESE DIFFERENT POSITIONS WILL RESULT IN VERY SMALL SHADOWS OR A BLACK DOT ON THE SCREEN.

PLEXIGLAS SCREEN

By tracking these position changes, you are mapping the direction of the rays of light that are diverging in all directions from the filament of the flashlight bulb.

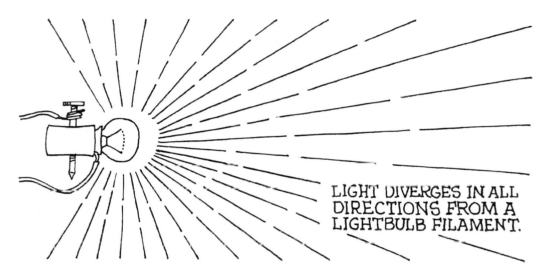

LIGHT DIVERGES IN ALL DIRECTIONS FROM A LIGHTBULB FILAMENT.

The results of the three preceding sets of activities are important. You can think of a shadow as a picture of an object, although a limited one. In the three sets of activities the shadows give you hints of what the real objects look like. You see mainly the outlines of the objects. If you rotate an object, you get more information about it, because you can see if the other sides of the object are the same or different. When first looking at the shadow of an object, you can't tell whether the object is three-dimensional or two-dimensional. The only way to find out is to rotate the object. This is less true of the wire shapes. Their shadows show clearly whether the wire shapes are two- or three-dimensional. You can tell whether the wire shapes are near or far away from the Plexiglas screen because the shadows are sharper when the wire shapes are nearer the screen.

Shadows of Transparent Objects

By a *shadow*, we usually mean the dark image on the screen, surrounded by light. The dark area shows where the object has blocked the light rays. From this observation it would seem that transparent objects would not make shadows. You can see out of windows and through glass jars or plastic soda bottles. How could they make shadows? Wouldn't the light be just a little weaker after passing through the glass or plastic? Not necessarily. When light encounters transparent objects, something strange happens. The object forms a pretty and interesting image very different from those seen in the previous sets of activities.

You will need:
▼ Shadow box and flashlight bulb arrangement from pages 29–30
▼ Collection of glass jars, soda bottles, and other empty glass and plastic materials, either clear or colored, with labels and glue thoroughly removed (Some suggestions are: mayonnaise jars, salad dressing jars, large or small cylindrical jars, cut-glass jars, or square jars; transparent covers from salad containers; plastic
▼ Bottles, such as those used for shampoo, baby oil, or dishwashing soap.)
▼ Package of 3-by-5-inch index cards
▼ Masking tape
▼ Microscope slide (optional)

Getting Started
All the transparent objects should be cleaned as thoroughly as possible. Dirt, food, or oil on the surfaces can confuse your observations.

Don't forget to make drawings of what you observe and record comments about what you did and what you saw. These observations will be helpful in the remaining activities.

Experiments to Try

SAFETY NOTE:
When using glass containers, be very careful not to drop them.

▼ Hold a container in the middle of the shadow box and see what results you get on the Plexiglas screen. Rotate the container so that light passes through the sides, bottom, and so forth. You will get both dark spots and bright spots inside the outlines of the image. Include both in your drawings.

▼ Start with a bottle from your collection of bottles and containers. Move it close to the light and observe the image on the Plexiglas screen. Slowly move the bottle toward the screen. As you do this carefully watch the image on the screen. At what distance from the light do you get the clearest image of this bottle? Make a measurement and record this in your notebook. Following the same procedure, try out all the other bottles and containers in your collection.

▼ Move one of the bottles up to the top edge of the Plexiglas screen and all the way back to the light. What happens to the pattern of light when you do this? Try this with all the other bottles and containers that you have collected.

▼ What is the difference in the images cast by plastic and glass bottles?

▼ What is the difference in the patterns made by the sides and the bottoms of containers?

▼ Is there any difference in the patterns made by square and round jars?

▼ Tape index cards to the sides of jars as shown in the illustration so that the edges are blocked. How does this change the resulting shadow?

▼ Make a deep scratch on a plastic container. Does this scratch show up in the

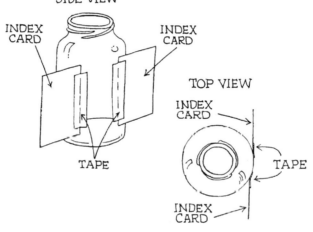

pattern on the Plexiglas screen? Does it make a difference if the object is close to the light or far away?

▼ If you have a microscope slide, make sure it is very clean. Hold it up to the light. Do you see anything on the Plexiglas screen?

▼ Place the slide at several different locations and map the images.

▼ Hold the microscope slide and rotate it as shown in the illustrations. What results do you see on the Plexiglas screen?

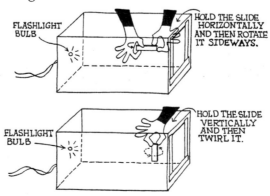

What's Happening?

Almost all of the transparent glass and plastic containers make some kind of shadow, but these shadows are different from the shadows in the previous sets of activities. They look more like shaded drawings. Within the outline of each image there are dark areas, gray areas, and bright spots of light. The glass bottles produce bright spots of light, some of them pleasing curved shapes. Some plastic containers produce gray images that are dark like regular shadows in some places. Thin plastic containers such as soda bottles usually make outlines with small dark marks here and there within the outline. Dark spots may show up where there is glue, dirt, or a scratch. When you make a big scratch on a container, it usually shows up on the Plexiglas screen as a line. The closer the bottle or container is to the screen, the sharper and more clearly the line will show up. When you tape index cards to a bottle or container, the shadow on the Plexiglas screen is just a scattered pattern of light and dark spots. The edges of the container don't show up. When you experiment with colored bottles, you should find that the results are similar to those you obtained with clear transparent bottles, except that the shadow will be the same color as the bottle. If the bottle is deeply tinted, letting little light through, it may be hard to see some of the effects.

Here are some examples of patterns obtained from a few bottles.

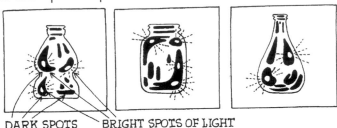

DARK SPOTS   BRIGHT SPOTS OF LIGHT

When light travels through the bottom or sides of some jars, especially thick ones, a pattern of bright spots will appear inside the dark and gray areas of the shadow. This also will happen when the light goes through the sides of containers such as shampoo or salad dressing bottles.

As you found in earlier experiments, when you move glass jars and some plastic bottles toward the Plexiglas screen the image will get smaller, The closer to the screen, the sharper and better defined the shadow will be. When the jars are right next to the Plexiglas screen you can sometimes read any raised lettering that may be on the bottle.

If you study closely the patterns created by the jars, especially where the bright spots show up in the shadow, and match these patterns to specific parts of the bottle, you have a clue as to why these patterns of light occur. Examine the glass to see how uniform the thickness is. If you experimented with a cut-glass jar or one with a raised surface, you will see that these patterns on the glass create patterns of light and shadow on the screen.

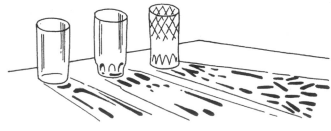

These observations show that curved surfaces, uneven surfaces, or raised lettering change the pathway of the light by bending the light rays. Depending on the particular pattern in the glass, the light rays may *converge* (be concentrated in one spot) or *diverge* (spread out). This is a basic scientific phenomenon. Light rays are bent when they go through one kind of material to another. The amount of bending depends on the material the light passes through. In this activity light is bent as it goes from the air into the glass, and it is bent again when it goes into the air again. This effect will be investigated further when you explore magnifying lenses later in the book.

If you experiment with a very clean microscope slide, you observe that the only thing you see on the screen is the edge of the glass, even if you rotate the slide sideways. This kind of glass is especially made so that it is very flat and evenly thin. You can confirm this by observing the image that appears on the Plexiglas screen as you rotate the slide. This image will be very even, with no curved lines.

There is another effect that you may notice when you experiment with the microscope slide. As you rotate the glass you see both a *reflection* (an image of bright light) and a shadow outline. If you watch the two as you rotate the slide, they will eventually merge. If you continue to rotate the slide, they will eventually merge. If you continue to rotate the slide, the reflected light grows larger and moves away from the dark line on the screen.

The point where the shadow and the reflection merge is the point where the surface of the glass is parallel to the direction of the light rays. When you place the slide at several different locations in the shadow box and find this merger point each time, the tilt of the glass will map the direction of the light rays.

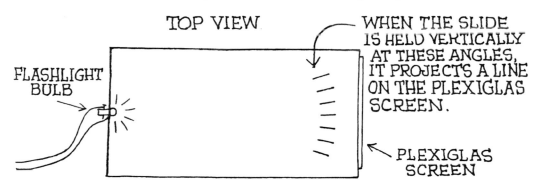

The angles at which you must turn the slide are similar to the results you got when you turned the piece of thin wire, the dowel, the comb, and the window screening in front of a light source.

Shadows of Thicker Transparent Objects

Most of the transparent objects you investigated in the previous activity were containers of one sort or another. You may have noticed that when you look through these containers, things are distorted. When they are filled with a liquid, they do strange things to light. Light shining through a glass of water will make bright spots on nearby surfaces. The water in swimming pools does something similar. It distorts the people or objects underwater. And on the bottom of the pools bright lines appear if the surface is disturbed.

The reason for these distortions, bright spots, and shadows is that water changes the pathway of light rays. So what would happen if you added water to the containers you used in the previous investigation? You can use the shadow box to explore this question. You may be surprised at the results.

You will need:
▼ Shadow box and flashlight bulb arrangement from pages 29–30
▼ Collection of transparent bottles (See page 41 for suggestions.)
▼ Hair comb
▼ Large bucket of clear water
▼ Food coloring – red, green, yellow (optional)

Getting Started

In this investigation, it is helpful to be more conscious of the shapes of the bottles. You should try to collect cylindrical bottles of several different diameters and bottles that are square, like instant-coffee containers. Salad dressing bottles will also be useful, because they have almost flat sides.

Make sure that the bottles are very clean inside and outside. All labels should be removed and any glue on the surface scraped away. Fill each of the bottles with water until they are about three-quarters full. One by one, place each of these in the shadow box. Move the container to different parts of the shadow box, and watch the image on the Plexiglas screen to get a sense of how the light changes.

Experiments to Try

SAFETY NOTE:
When using glass containers, be very careful not to drop them.

Place a bottle right in front of the light, and observe the image on the Plexiglas screen. Then very slowly move the bottle toward the Plexiglas screen. As you do so, observe how the image changes. Do this with all the bottles.

▼ As you move each bottle, at what point do you see a bright, narrow vertical line on the screen?

▼ At what point do you see a good outline of the bottle on the screen?

▼ At what point do you see a wide beam of light coming from the bottle?

▼ Does the light behave differently when it goes through wide and narrow bottles?

▼ Does the light behave differently when it goes through narrow-sided salad dressing bottles versus round bottles?

▼ What happens when food coloring is added to the water? How does this change the image on the screen?

▼ Place a bottle filled with clear water a few inches away from the light. Hold a comb between the light and the bottle. What happens to the image on the Plexiglas screen when you move the comb close to the light, and then close to the bottle?

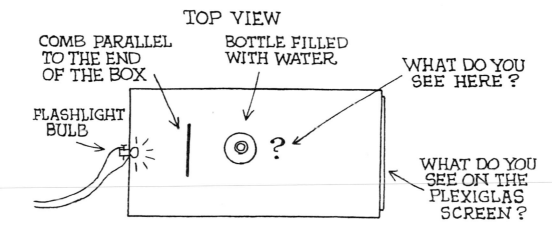

▼ Place the bottle filled with clear water at each of the five different positions shown in the illustration.

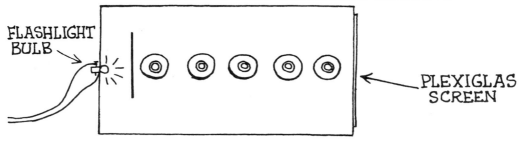

Move the comb back and forth between the light and the bottle. How does the shadow of the comb's teeth change as you try these different locations? Repeat the whole experiment by putting the comb between the bottle and the Plexiglas screen. What happens to the shadow of the comb's teeth as you move the bottle?

What's Happening?

The water in the container changes the way the light travels through them. As you move each container away from the light toward the Plexiglas screen the image on the screen changes. When the bottle is very close to the light, the shadow is spread out and the image is fuzzy. A few inches from the light the bottle full of water makes a bright, narrow vertical line on the Plexiglas screen. Moving it farther from the light makes the bright line disappear and gives a dimmer light again. This outline may be fuzzy or sharp, depending on the kind of bottle you are using. (For example, plastic containers that are used for food will give fuzzy images, while some new plastic bottles that do not have dust or scratches on the surface will give clear outlines.) When the bottle is just a few inches from the screen a vertical bright line appears again.

So, the image of the bottle projected onto the Plexiglas screen will vary depending on where you place the bottle in relation to the flashlight bulb and the Plexiglas screen. The position where the shadow looks most like a drawing of a real bottle is about halfway between the light and the screen. As with empty bottles, those that have raised or uneven surfaces also make dark sections and bright patterns of light within the outline of the shadow.

Adding food coloring to the water in the bottles gives pretty results: The parts of the shadow that were previously white are now the color that you added. However, everything else about the image remains the same as it was when the bottle was full of plain water.

The shadows cast by the containers of water give more information about the bottles than the shadows produced by the opaque, or solid, objects in previous investigations. They show the curves and lines on the surfaces of the bottles. Artists use this fact in depicting curved objects. They add shading to different parts of their pictures to show where there is curvature.

The results you obtain are true for most bottles. If the bottle is quite large or is etched with special designs, the image on the screen will be harder to see.

The position at which the bright, narrow vertical line appears on the Plexiglas screen depends on the diameter and shape of the bottle. The larger the diameter, the farther away from either the light or the Plexiglas screen the bottle must be before the line appears. The smaller the diameter, the closer the bottle must be to the light or the Plexiglas screen before the line appears. Bottles with almost flat surfaces, such as salad dressing containers, give a slightly different result. They bend the light to a vertical line, but the line is wider than those obtained with cylindrical bottles and is more divergent. Also, the change is gradual. In cylindrical bottles, this change happens rapidly as the bottle is moved.

If you examine carefully the light beam between the bottle and the screen, you can begin to understand what is happening. As you move a cylindrical bottle away from the light, toward the Plexiglas screen, there is a spot where the light in front of the bottle meets at a point on the bottom of the shadow box. The light coming from the bottle is bent so that all the rays are converging at that point. If you look beyond this spot, you can see that the light diverges again, making a pattern like a big X.

TOP VIEW

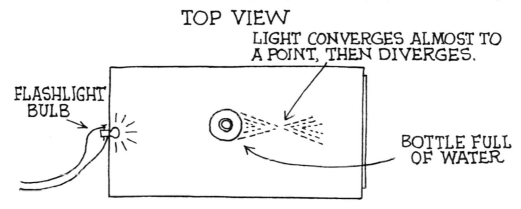

To observe this further, it is helpful to use a comb. Look at the illustrations below. When the bottle and comb are at Position A, the light spreads out from the comb. The line shadow of the comb's teeth will show divergent rays.

TOP VIEW

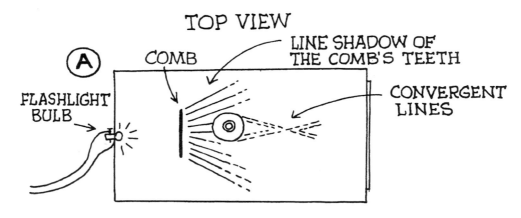

When the bottle is at Position B, near the Plexiglas screen, the line shadow of the comb's teeth shows convergent lines.

TOP VIEW

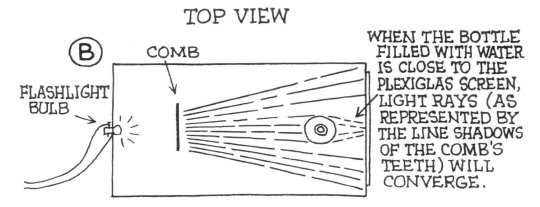

WHEN THE BOTTLE FILLED WITH WATER IS CLOSE TO THE PLEXIGLAS SCREEN, LIGHT RAYS (AS REPRESENTED BY THE LINE SHADOWS OF THE COMB'S TEETH) WILL CONVERGE.

There are divergent and convergent lines in both situations, A and B. Where the divergent lines meet the bottle, they emerge from the bottle as convergent lines.

This is true for a jar that is about three or four inches in diameter. The same changes in the shadow of the comb's teeth will occur with other cylindrical jars, but at different distances.

Each of the containers acts like a lens or magnifying glass. It bends the light, concentrating the light rays at one point or spreading them out (having them diverge) at another point. In a later activity you will investigate a similar phenomenon with real lenses.

Cameras

In all the previous activities, you made shadows by placing objects between a light source and an opaque surface. The resulting images sometimes looked like black-and-white pencil drawings of the objects. With transparent objects, shadings of grays and black areas appeared, making the resulting picture more realistic and also more three-dimensional. Except when you added food coloring to the water in the jar or used a colored container, the pictures didn't have any color. Nor did you have a sense of the surface of each side of the object. An opaque object could have had indentations and scratches. These would not show up in the shadow on the Plexiglas screen. For instance, if you were to put indentations in the Plasticine, these markings wouldn't show up in the shadow.

In this next set of activities, you will use cameras to make images that look much more like the original object. One big difference between these activities and the earlier ones is that now the object will be outside the box.

Experimenting with a Box Camera

For thousands of years, artists have attempted to show an object as it actually appears to our eyes. They have invented a number of ways of doing this. One invention was based on a discovery made a long time ago but not fully used until the past few hundred years. Centuries ago, ancient Greeks and, later, Arabic scientists recorded seeing a picture appear on a wall in a dark room if there was a very small hole in the opposite wall. Light reflected off the outside object came through the hole and formed an image of the object on the flat surface of the wall. The image was not a shadow, but a picture showing the object with its real colors and surfaces. The room in which this occurred had to be very dark and the hole of a small size.

During the fourteenth, fifteenth, and sixteenth centuries, artists and scientists were quite curious about this phenomenon and carried out many kinds of experiments to explore it. Artists studied these images to help them draw and paint scenes and objects as they ordinarily appear to us in life. Scientists used these images to learn about the properties of light and how images form in our eyes.

This room-sized image making even became a form of entertainment. In Victorian England and some parts of America, large rooms were built with a tiny hole in one wall. If the rooms were very dark, an entire outdoor scene was seen on the wall opposite the hole.

The rooms in which experiments of this type were done were called *cameras*, from the Latin word for "chambers." This is the origin of the word we now use for a device that takes pictures.

Before the modern camera there were several different versions. You can make some models of early cameras with simple materials. By doing some experiments, you can learn more about picture making and about the properties of light.

The Box Camera

To make a box camera you can use some of the materials from earlier activities. The picture this camera produces will look like the real object, with color and surface texture. It will be very different from the black shadows you saw in earlier experiments.

You will need:

▼ Large cardboard box, the bigger the better (Refrigerator cartons are the best, but any box large enough to but your head and arms into is okay. It it's big enough, reuse the shadow box from pages 23–24 with the Plexiglas screen and bulb holder removed.)
▼ Heavy blanket
▼ Piece of Plexiglas or other transparent plastic from page 23
▼ Pad of tracing paper, 9 inches wide and 12 inches long
▼ Duct tape, at least 2 inches wide
▼ 2 large paper clips
▼ Pencil
▼ 2 or 3 paper plates, 10 inches in diameter
▼ Ruler
▼ Compass
▼ Brass fastener, 1 inch long
▼ Utility knife
▼ Coins – a penny, nickel, dime, and quarter
▼ Nail

Getting Started

Step 1. Lay the box on its side so that the flap ends are free. Tape one flap end firmly closed with duct tape. The other flap end will be the doorway.

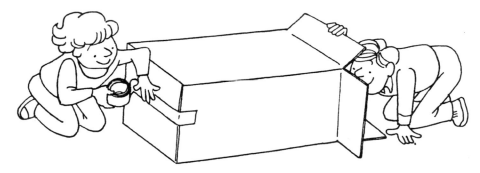

Step 2. During the experiments, the box must be completely dark. Since the doorway end of the box will have some open slots or holes, you will have to have a friend cover them from the outside. Tape a heavy blanket around the doorway and the end of the box to use as a cover.

If you are using the shadow box from the previous activities, tape the blanket over the end of the box where the Plexiglas screen was attached.

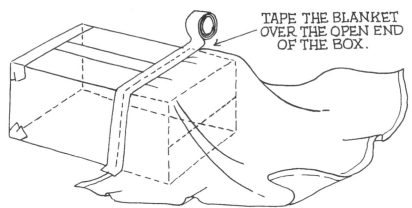

TAPE THE BLANKET OVER THE OPEN END OF THE BOX.

Step 3. Check the box to make sure there are no holes anywhere. This is easiest to do from the inside, so stick your head in or crawl inside and look for places where you see light. Patch these holes with the duct tape.

Step 4. Use the knife to cut an opening about two inches square in the middle of the flap end opposite the doorway. The edges of the opening should be as clean as possible.

Step 5. Tape two or three paper plates together to form a thicker circle. Find the center of the group of plates by measuring carefully with a ruler, or by balancing the plates on the flat eraser head of a pencil. The plates will remain horizontal when the eraser is held in the middle of the plates. Make a mark where the eraser touches the plates.

Step 6. In the center of the top plate, use a compass to measure a circle that has a 2-1/2-inch radius. Place a quarter, a nickel, a penny, and a dime on the circle, lining up each coin so that its center is on the line of the circle. Position the coins so that they are about three inches apart. Trace around the edge of each coin to obtain different circles of different diameters.

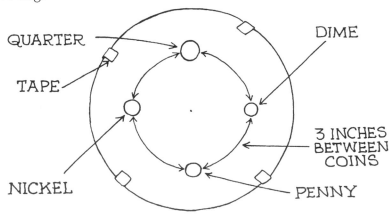

QUARTER

DIME

TAPE

NICKEL

3 INCHES BETWEEN COINS

PENNY

Step 7. Carefully cut out the outlines of the coins.

Step 8. Punch a hole in the middle of the plates with a nail.

Step 9. Position the plates on the outside of the box so that one of the coin-sized holes in the plates lines up with the two-inch hole in the cardboard box. Rotate the plates to make sure you can line up each hole with the opening in the box.

Step 10. Anchor the plates to the outside of the box with the brass fastener. Use the nail to make a hole in the box so that the fastener slips through easily.

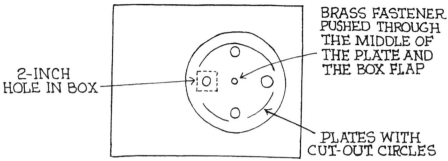

FRONT END OF BOX CAMERA

Setting Up

Step 1. It is best to set up the box camera outdoors on a bright sunny day. If the weather does not cooperate, you can try setting it up indoors in a large room that is brightly lit or facing a window.

Step 2. Attach a piece of tracing paper to the piece of Plexiglas with paper clips. This will be the screen for viewing the picture inside the box. You will be holding it in your hands and moving it around.

Step 3. If you are working with a small box, you will have to put your head and arms in the box while holding the Plexiglas screen and a pencil in your hands. If you're using a refrigerator box, crawl inside with the Plexiglas screen and a pencil.

Step 4. Ask your friend to close up the opening with the blanket so that it is completely dark inside and no light leaks in from that end. Hold the Plexiglas screen up to the light coming through the hole in the box. Move the Plexiglas screen back and forth toward the light until you see a clear picture on the screen. Trace the image on the tracing paper.

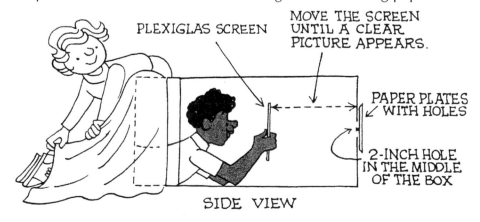

SIDE VIEW

Experiments to Try

▼ As you look at each object or scene, have your friend rotate the plate so that you can experiment with each of the holes you cut out. How does the image change as the holes get larger? How does the intensity of light on the Plexiglas screen change as the holes get larger?

▼ Ask your friend to move the box camera so that it aims at different things, such as nearby cars, trees, and houses. How does the image change as you go from bright areas to dark ones?

▼ Have your friend stand in front of the box camera so that that his or her picture appears on the Plexiglas screen inside the box. What happens to the image as your friend backs away from the box?

▼ Try setting up the box camera in your house and aiming it out a window. Can you see an image of something outside on the Plexiglas screen?

▼ Try looking at an object in the room, something that has both dark and light areas. Does an image still appear on the screen?

What's Happening?

The sharpness of the image depends on the size of the hole the light passes through. As the hole gets bigger the image gets fuzzier until there is no image at all. The smaller the hole, the sharper the image. In your box camera, images formed through a *very* small hole will be very dim and hard to see. The brightness of the image depends on how bright it is outside the box camera. Outdoor scenes on bright sunny days will show up well on the Plexiglas screen. If the sun goes behind the clouds, the image will fade but still be visible. Looking at scenes indoors gives similar results. Where there are bright lights an image will appear on the screen, but areas that are dark or barely lit will not even show up. Any extra light entering the box camera will dim the image, which is why you have to cover every hole.

The size of the image depends on the distance between the object and the Plexiglas screen. Inside the box camera, you can make the image bigger by moving the Plexiglas screen farther from the hole. Or outside the box camera, you can move the object closer. Objects that are far away will look small on the screen, while nearby ones will look bigger. Therefore, as a person backs away from the box camera, his or her image becomes smaller.

You may have been surprised that the image on the Plexiglas screen is upside down, no matter what the object is. This is why: Light is coming from the object to the hole in the box camera. The light from the original source, such as the sun or very bright indoor fixtures, is absorbed by the object then reemitted in divergent rays.

Of these reemitted rays, the only ones that enter the box camera are the ones aimed right at the hole. Other rays are blocked. If you trace the path of light emitted by a person or object into the hole, you can see why the image is upside down.

Light rays coming from the top of the object go straight through the hole to the bottom of the Plexiglas screen. Light rays coming from the bottom of the object go through the hole to the top of the screen. The ones from the middle of the object go to the middle of the screen.

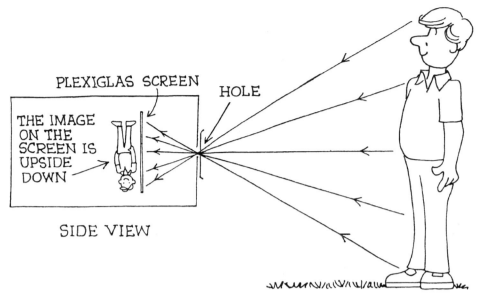

With the box camera you get a real image of an object, showing all its colors and surfaces. It is much more revealing than the dark shadows that were produced in previous activities. This type of camera was a popular device with artists and scientists for many years. Many variations were built so that different kinds of pictures could be made.

Improving the Box Camera

Although you can get an image with the simple box camera, the picture is faint. People discovered that they could get a sharper, brighter image if the light passed into the camera through a lens instead of a simple hole.

The first device that used lenses like this was called a *camera lucida*. It was the forerunner of the modern camera we use today. With just a few extra materials you can make one of these in a primitive form and investigate how it works.

In this activity you use magnifying glasses as lenses. You can find them in many department stores, science-museum gift shops, and hobby shops. You can use magnifying glasses with or without handles.

You will need:
▼ Box camera from pages 52–54, with the paper plates removed
▼ Plexiglas screen from page 23
▼ Masking tape
▼ 1-inch-diameter magnifying glass
▼ 2-1/2-inch-diameter magnifying glass
▼ 3-1/2-inch-diameter magnifying glass
▼ 3 square pieces of heavy cardboard, six inches on each side
▼ Utility knife
▼ Pencil
▼ Ruler

Getting Started

SAFETY NOTE:

Be careful! In these activities with lenses use only tracing paper. Under certain conditions other kinds of paper, and other materials, may smolder or burn.

Step 1. With a knife, enlarge the two-inch opening in the box camera to a circular opening approximately four inches in diameter.

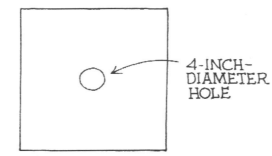

FRONT OF BOX CAMERA

Step 2. Place a magnifying glass on a piece of cardboard. Trace around the edge of the magnifying glass. Cut out the outline. Tape this magnifying glass onto the piece of cardboard so that the magnifying glass fits into the hole. Repeat these steps for each of the magnifying glasses so that each one is mounted on a piece of cardboard. These are the camera lenses.

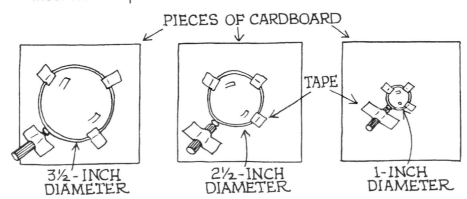

Step 3. Tape a mounted camera lens inside the box camera so that the magnifying glass is over the hole in the box.

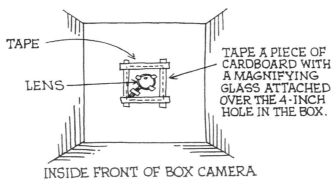

INSIDE FRONT OF BOX CAMERA

Step 4. Aim the box camera at an object or person.

Step 5. Put your head and arms in the box and hold the Plexiglas screen in your hands. Have your friend use the blanket to block any light leaking into the box camera. Hold the Plexiglas screen up to the light coming through the hole in the box. Move the Plexiglas screen back and forth toward the light until you see a clear picture on the screen. You can even try tracing the image on the tracing paper.

When making an image, look carefully at the picture produced to see if the whole picture is clear and focused.

Experiments to Try

When doing the following experiments, pick one object that is well illuminated. Point the magnifying lens at this object, using it as your test subject. Use this object for making comparisons among different kinds of lenses. One by one, experiment with all the lenses.

▼ Where do you have to place the Plexiglas screen to get a sharp image when you test the different lenses? Is the Plexiglas screen the same distance from the lens each time?

▼ Are there any differences in the size of the images produced by the different lenses?

▼ Put two or more lenses together. Begin by holding them right next to each other. Then move them several inches apart. What happens when two or more lenses are combined? If an image is produced, how does it differ from those made when the lenses are used separately?

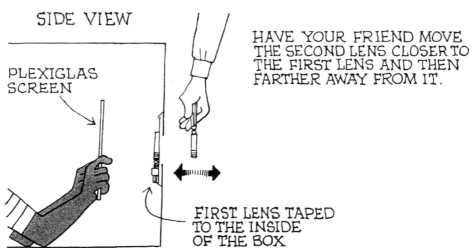

What's Happening?

The results you obtain vary with the size and the shape of the magnifying-glass lenses. Larger lenses produce larger pictures that are in focus farther from the lens. However, if you look very carefully at the total picture on the Plexiglas screen, you should notice that the pictures from the larger lenses tend to be a little blurry around the edges, especially when compared to the pictures produced by the one-inch lens.

If the lens is in direct sunlight and you move the Plexiglas screen back and forth, you will find a position where there is a bright spot of light on the tracing paper. All of the light rays striking the lens are being concentrated at this spot. If you hold the Plexiglas screen right next to the lens and slowly move it away, you can see how the light rays are being concentrated to a place where a good image occurs. As you continue to move the Plexiglas screen farther away from the lens, the light diverges again and the image looks fuzzier. These results are similar to the results you obtained from your experiments with the bottles filled with water on page 46. Lenses and bottles filled with liquid both bend the light rays and focus them to make an image. Therefore, you can think of a bottle filled with water as a giant but crude lens.

This last observation helps explain how a lens functions. Recall what happened in the experiments on page 42, when a microscope slide was placed in front of the flashlight bulb. If the glass was very smooth and of high quality, no bright or dark spots appeared on the Plexiglas screen. The light rays seemed to go through the glass undisturbed. This isn't always true for ordinary glass, which tends to be a little bumpy and uneven so that when light passes through, it is bent by the imperfections in the glass. However, even with a high-quality glass such as a microscope slide, there is a sideways displacement of the light when it hits the slide at an angle.

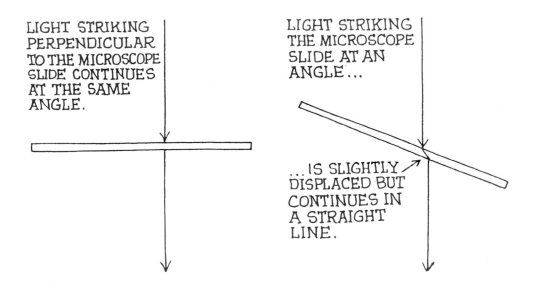

LIGHT STRIKING PERPENDICULAR TO THE MICROSCOPE SLIDE CONTINUES AT THE SAME ANGLE.

LIGHT STRIKING THE MICROSCOPE SLIDE AT AN ANGLE...

...IS SLIGHTLY DISPLACED BUT CONTINUES IN A STRAIGHT LINE.

This displacement isn't usually noticeable with thin glass, but it is more apparent with very thick glass such as the kind used for entrances to stores or banks. Viewing your hand behind a very thick glass will give strange results. Viewed straight ahead, it will look normal. When looked at from an angle, however, the hand will appear displaced. Objects viewed through the edges of thick glass also appear to be displaced.

In a lens, the bending of the light can be controlled by the material used to make the lens and by the size and shape of the lens itself. Most of the lenses you used in this set of experiments were probably made from the same type of plastic, and therefore bent the rays in the same way. Because the lenses were of different sizes and shapes, they focused the image or the spot of light at different distances from the lens.

One way of making a comparison among lenses is to imagine what a section of lens would look like. If one of the plastic lenses were cut in half, it would look like the middle drawing below.

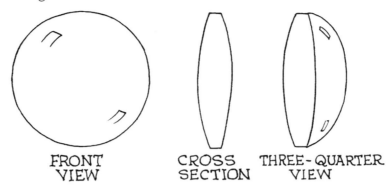

FRONT
VIEW

CROSS
SECTION

THREE-QUARTER
VIEW

This is similar to cutting an orange in half. Because the lenses you used were of different diameters, their cross sections would show that their centers have different thicknesses and that they taper to the edges differently. These differences cause the light to bend and focus at different distances from the lenses.

You can now consider what happens to the rays of light when they move through the cross section of the lens. Recall the shadow lines you saw when the comb was placed in the shadow box with the bottle filled with water. (See pages 48-49.) When these shadow lines go through a bottle or lens, they can be seen to converge to a point. With the bottle this point of convergence is the same point where a bright line is produced. This point's special scientific name is the *focal point*. With the lens, a bright spot of light is produced at the focal point. If you looked carefully at the point of convergence produced with the comb's shadow, you could see the shadow lines diverging or spreading out again, beyond this point. Now recall your experience of moving the Plexiglas screen away from the box camera lens. At first there is a fuzzy light, then a bright spot, and then a clear picture not too far from the bright spot. The light rays from the object form a small picture of the object inside the box. This image forms behind the concentrated spot of light. The distances will vary depending on the size and the shape of the lens.

If the lenses were perfect, the light rays from the object would meet exactly at the point where the bright spot forms, and they would make a very small spot. However, most lenses do not allow this to happen. The result is that the picture that forms can be slightly fuzzy. This is particularly noticeable with cheap large magnifying lenses. The edges of the image produced by these lenses are fuzzy and even blurred.

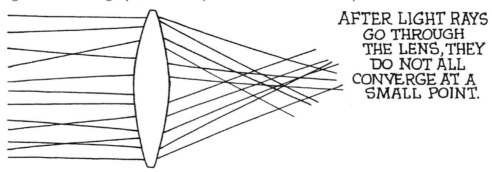

AFTER LIGHT RAYS
GO THROUGH
THE LENS, THEY
DO NOT ALL
CONVERGE AT A
SMALL POINT.

You also may have noticed when working with the bottles filled with water and with some of the lenses that parts of the picture show bits of rainbows or colored light. Even very good lenses have this problem. White light contains all the colors of the rainbow, among them red, blue, and green. As the white light goes through the lens it can be slightly separated into these colored lights.

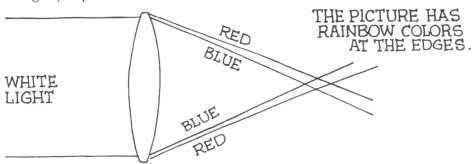

Both of these problems can be corrected by using a combination of lenses. Did you notice that when you used two lenses in your experiments a much sharper image was produced?

If you cut the lens cylinder of a very expensive camera in half, you would see not one lens but a series of several lenses of different shapes. These are used to overcome the fuzzy images and the separation of the white light.

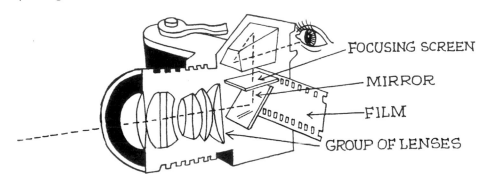

Cheap cameras generally use a single lens. This produces a satisfactory picture for most people. However, professional photographers usually want very sharp pictures. Therefore, they use cameras that have a combination of several different lenses.

Improving the Picture from Lenses

Even high-quality lenses produce slightly distorted pictures. Adding lenses helps overcome this problem, but so does another technique: directing the light through the middle of the lens. You can see how this happens by experimenting with the lenses used in the previous activity.

You will need:
▼ Box camera and three mounted camera lenses from page 56–58
▼ Group of paper plates with different-sized holes from pages 53–54
▼ Plexiglas screen from page 23
▼ Masking tape

Getting Started

SAFETY NOTE:

Be careful! In these activities with lenses use only tracing paper. Under certain conditions other kinds of paper, and other materials, may smolder or burn.

Step 1. Set up the box camera either outdoors or indoors facing a window.

Step 2. Tape a mounted camera lens over the hole in the side of the box camera. (See pages 57–58 for assembly directions.)

Step 3. Holding the Plexiglas screen, put your head and arms inside the box and have your friend close the doorway with the blanket. Hold the Plexiglas screen up to the light coming through the hole in the box. Move the Plexiglas screen until the image is as sharp as you can make it. Then move the Plexiglas screen again so that the image is slightly out of focus. Have your friend hold the group of paper plates a few inches in front of the box camera so that light goes through one of the holes in the paper plates and then through the lens. Holding the hole in the paper plates a few inches from the lens will improve the quality of the picture on the Plexiglas screen.

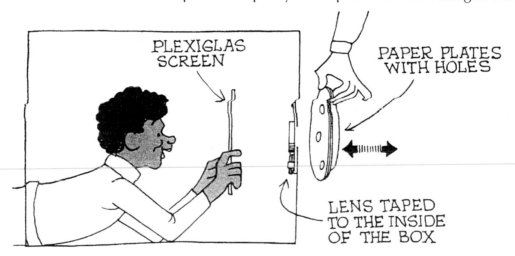

Experiments to Try

▼ With the same lens on the box camera, have your friend hold different-sized holes in front of the lens. What happens to the image as the holes get bigger? Smaller?

▼ Test all the mounted lenses. How does the image change as the lens size changes?

▼ What happens if you match the largest lens with the largest hole? The smallest lens with the smallest hole? The largest lens with the smallest hole, and vice versa?

What's Happening?

As the hole in front of any of the lenses is made smaller the picture quality improves, but it also becomes dimmer. Recall what happened when you were just using the holes in the box instead of a lens. Smaller holes produced a sharper picture but also a dimmer one, because the holes in front of the lens select only a few of the rays coming from the object.

Larger lenses distort more that smaller ones. As you saw in your experiments, the image of the picture is improved, particularly around the edges, when you hold a paper plate with a hole in it in front of this kind of lens. However, when you start matching up the smallest hole with the biggest lens, you start seeing only part of the full image, and it will get dimmer, With a small hole, part of the entire picture is cut off, so you see less of an object than you do when viewing it through a larger hole. The biggest improvement, or the sharpest picture, will be with the smallest hole in front of the smallest lens. However, unless your box camera is very dark inside, these improvements may be hard to see.

There is one interesting and useful observation you may have made, When you move the Plexiglas screen back and forth in front of the lens by itself, the place where the best image occurs is fairly precise. When you add the paper plates so that the light comes through a hole to the lens, you can move the screen back and forth a little and still have a good picture. This arrangement is used in expensive cameras. In addition to a series of lenses, there is a device that functions like the holes in the paper plates. It is called a *diaphragm*, and it creates holes of different diameters near the camera lens. By using these different openings, a photographer can take one photo and get sharp images of both nearby objects and those far away.

A Further Exploration

You learned earlier that light is reflected from objects. Now that you have learned about lenses, you can experiment with this phenomenon.

You will need:

▼ Slide projector
▼ 2-1/2- or 3-1/2-inch-diameter magnifying glass (See page 56 for suggestions on where to purchase this.)
▼ Full-color picture from a magazine
▼ Piece of unlined white paper, 8-1/2 inches wide and 11 inches long

Step 1. Line up the slide projector and the picture so that light shines on the picture at a slant. Hold the magnifying glass in front of the picture and try to get an image of the picture on the paper.

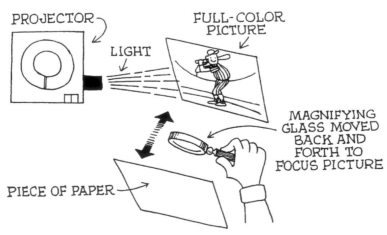

PROJECTOR

LIGHT

FULL-COLOR PICTURE

MAGNIFYING GLASS MOVED BACK AND FORTH TO FOCUS PICTURE

PIECE OF PAPER

You'll notice that the image is upside down.

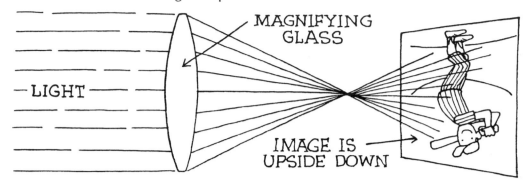

What's Happening?

The light from the projector is absorbed by and given back out in the colors of the picture. Lining up the magnifying glass at the correct angle focuses these light rays on the surface of the paper.

The intense, focused light from the projector allows you to project an image. Sunlight could be used too if it were directed into a dark box or room where an object was to be illuminated.

The arrangement you used here is very similar to an opaque projector, illustrated below, which has very powerful lights. By combing these lights with the right combination of lenses, you can use this type of projector project a large image of a picture on a screen.

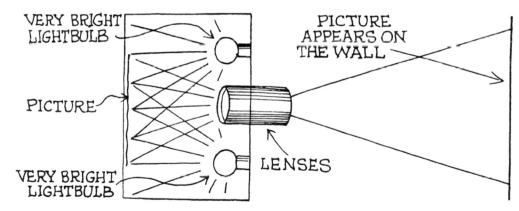

Experimenting with a Real Camera

The ability to produce a small image of a person or a scene was possible several hundred years ago, although it wasn't a photograph in the way we think of it today. Then several inventors came up with the formula for light-sensitive paper – the first film – so it was possible to make a permanent record of what the camera saw. In this activity you'll experiment with a real camera and take real pictures.

You will need:
▼ Disposable camera, or a conventional camera
▼ Magnifying glass from page 56
▼ Package of 3" x 5" index cards
▼ Scissors

Step 1. Buy a disposable camera that can take outdoor pictures. There are several brands on the market, available in drugstores and department stores. (If you prefer, you can use a conventional camera, with ASA 100 film.)

Step 2. Read the directions so you know how to work the disposable camera.

Step 3. Cut out several different-sized holes in the index cards. Each hole should be smaller than the diameter of the magnifying-glass lens.

Experiments to Try

As you take each picture, record the negative number (it should appear in a small window on the camera). Also record how you took the picture.

▼ Take a picture of your friend riding past you on a bicycle or running in front of the camera. Ask your friend to ride or walk past you two or three more times at different speeds. What kind of pictures do you get?

▼ What kind of picture do you get if you try to photograph a fast-moving car on a highway?

▼ What kind of picture do you get if the person or object you're photographing is in the shade of a tree or building?

▼ What kind of picture do you get if the sun is behind you and the person or object you're taking the picture of is in the sunlight?

▼ What if you are facing the sun when you take the picture?

▼ How does the size of a person in a photograph change as the person moves farther and farther from the camera?

▼ At what distance does the person's body fill the picture from bottom to top?

▼ Take a picture from the top of a tall building. How do the farthest-away objects look? The closest?

▼ Shooting through a magnifying glass, take a picture of your friend standing 10 feet away outdoors in bright sunlight. Then take several more photos of the same scene, each time putting an index card with a different-sized hole in front of the magnifying glass.

What's Happening?

The image you get on your photograph depends partly on the camera, partly on the light conditions at the time you take the picture, and partly on the light sensitivity of the film. Simple throwaway cameras have only a plastic lens and a shutter that opens and closes to let light reach the film. The lens is positioned at a specific distance in the camera so that a clear image falls on the film when the shutter is opened. The longer the shutter is open, the longer the film is exposed to light. If the film is *overexposed*, meaning that too much light enters the camera, the picture will be too bright. The image will be *underexposed* – dark, or even black – if the shutter was open too short a time. In simple cameras the shutter speed has been preset to give the right exposure for ordinary lighting conditions.

Light conditions are important too. If you point simple cameras at the sun, the resulting picture will look pale, because the film was overexposed. On the other hand, if you try to take a picture of an object or a person in shade, or during overcast weather, or close to sunset, the picture will look dark, because the film was underexposed.

Today film can be made of either paper or plastic. It is coated with chemicals that are sensitive to light. Different kinds of film are sensitive to different kinds, and different amounts of light. So some films produce black-and-white images rather than color. And some films work best in dim light, while others are best outdoors in full sun.

The size of the object in the picture depends on how close you are to it. The subject of the picture has to be several feet from the camera before you can capture its full height. Scenes of mountains, hills, or large open areas appear to be very small. Only objects that are near the camera will be noticeable and in good focus. Things that are far way from the lens of a simple camera will not be well focused. With some cameras this is also true for objects that are very close to the lens.

When you take pictures of moving objects, they may end up looking blurred in the picture. To take pictures of a fast-moving person or vehicle, the opening of the shutter of the camera has to be adjusted. The faster your subject is moving, the faster the shutter has to be opened and closed. The shutter in a cheap camera is present for stationary or very slow-moving objects.

What you can learn from these experiments is that you continually need to be thinking about several factors when taking a picture. The lighting conditions have to be within a certain range for the best results. You need to look through the viewfinder each time to make sure you have properly framed your subject. To take a well-focused picture of a faraway object, you need a special lens called a *telephoto lens*.

The Camera and the Eye

In these activities you learned a great deal about the properties of light. You saw how a camera, lens, and film can make pictures of the real world. Now you can use what you've learned to understand a much more complicated picture taker, the human eye. Parts of the camera and parts of the eye work in very similar ways.

A camera is a dark container. Light enters only through the lens system, which directs the image to the back of the camera. Cameras combine lenses in order to get a good picture. Different kinds of lenses have to be used in order to take pictures of people or objects at a short distance from the camera or very far away from it. Cameras also use a diaphragm (see page 63), which makes the light go through the middle of the lens, giving a sharper image. The light falls on the light-sensitive film, which captures the image and makes the picture.

The human eye is not exactly like a camera. For one thing, it is round, and filled with a Jell-O-like material that helps the eye hold its shape. Like a camera, it is a dark container, and light enters only through a lens. In the eye there is only one lens, but the shape of this lens can be changed by the muscles around the lens, called the *ciliary muscles*.

These muscles make the lens thick or thin. To see an object far away, the muscles relax to form a thin lens, which brings the object into focus. To see an object up close, the muscles contract, or become tense, which thickens the lens and brings the object into focus.

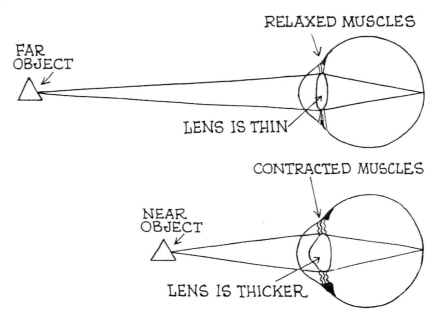

In the camera experiments, you shot through index cards having holes of different sizes. You found that the smaller the opening, the sharper the image. The eye can also change the size of its light opening, called the *pupil*. You can see this happen. Stand in front of a mirror and look at the black circle in the center of your eye. This is the pupil. Watch it as you shine a light in your eye. The *iris* – the colored part of the eye – closes the pupil so that less light enters the eye.

When the pupil becomes smaller, it produces an image that is sharper, but dimmer. The back of your eye, called the *retina*, compensates for this. The retina functions something like the film in the camera.

The surface of the retina is covered with tiny cells called *rods* and *cones*. The rods are sensitive to dim light, black, and white. The cones are sensitive to bright light and color.

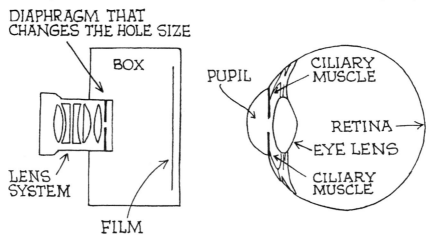

The rods and cones function differently in different light conditions. If you walk from the bright outdoors into a dark room, objects are hard to see at first, because the rods and cones have not adjusted. When they do, you can see what's in the room. (A camera has no retina. When the light conditions change dramatically, you have to use a different film.)

In your experiments with shadows, you found that a single view of an object often does not provide much information. You had to turn the shape or object and see it from different angles to know what it was. When you look at an object, you use both of your eyes, which helps you see two slightly different views of the object and gives you more information about how far away a thing is. You also move your head and body so you see the object in different lights and at different angles. In those ways you have a better understanding of what you are looking at.

So a camera is like your eye, but your eye is much more complicated and can do things a camera cannot.

Exploring Light, Exploring Lenses

Each of the investigations that you carried out in this book can be taken further. For instance, you can measure precisely the relationship between the size of a shadow and its distance from an artificial light source. Experiments like this would help you understand a branch of mathematics called geometry. It is the study of the properties and measurements of solids, two-dimensional objects, lines, and points in space.

The investigations that you carried out with magnifying-glass lenses were only a beginning. There are many more investigations that can be done to help you better understand how lenses work. For instance, the box camera can be modified in a number of ways to produce interesting distorted pictures. You can try to figure out how these distortions come about. You can even try to invent your own camera to see how different kinds of lenses change the final picture. You can also make your own projectors, telescopes, or microscopes. Each of these projects would involve using lenses in special ways – and would help you know more about making pictures and about the properties of light.